Symbolic Execution and Quantitative Reasoning

Applications to Software Safety and Security

Synthesis Lectures on Software Engineering

Editor
Luciano Baresi, *Politecnico di Milano*

The Synthesis Lectures on Software Engineering series publishes short books (75-125 pages) on conceiving, specifying, architecting, designing, implementing, managing, measuring, analyzing, validating, and verifying complex software systems. The goal is to provide both focused monographs on the different phases of the software process and detailed presentations of frontier topics. Premier software engineering conferences, such as ICSE, ESEC/FSE, and ASE will help shape the purview of the series and make it evolve.

Symbolic Execution and Quantitative Reasoning:

Applications to Software Safety and Security

Corina S. Păsăreanu

ISBN: 978-3-031-01423-9   paperback
ISBN: 978-3-031-02551-8   ebook
ISBN: 978-3-031-00341-7   hardcover

DOI 10.1007/978-3-031-02551-8

A Publication in the Springer series
SYNTHESIS LECTURES ON SOFTWARE ENGINEERING

Lecture #6
Series Editor: Luciano Baresi, *Politecnico di Milano*
Series ISSN
Print 2328-3319   Electronic 2328-3327

Symbolic Execution and Quantitative Reasoning

Applications to Software Safety and Security

Corina S. Păsăreanu
NASA Ames Research Center

SYNTHESIS LECTURES ON SOFTWARE ENGINEERING #6

ABSTRACT

This book reviews recent advances in symbolic execution and its probabilistic variant and discusses how they can be used to ensure the safety and security of software systems. Symbolic execution is a systematic program analysis technique which explores multiple program behaviors all at once by collecting and solving symbolic constraints collected from the branching conditions in the program. The obtained solutions can be used as test inputs that execute feasible program paths. Symbolic execution has found many applications in various domains, such as security, smartphone applications, operating systems, databases, and more recently deep neural networks, uncovering subtle errors and unknown vulnerabilities. We review here how the technique has also been extended to reason about algorithmic complexity and resource consumption.

Furthermore, symbolic execution has been recently extended with probabilistic reasoning, allowing one to reason about quantitative properties of software systems. The approach computes the conditions to reach target program events of interest and uses model counting to quantify the fraction of the input domain satisfying these conditions thus computing the probability of event occurrence. This probabilistic information can be used for example to compute the reliability of an aircraft controller under different wind conditions (modeled probabilistically) or to quantify the leakage of sensitive data in a software system, using information theory metrics such as Shannon entropy.

This book is intended for students and software engineers who are interested in advanced techniques for testing and verifying software systems.

KEYWORDS

symbolic execution, testing, constraint solving, algorithmic complexity analysis, probabilistic reasoning, model counting, software reliability, side-channel analysis

Contents

Acknowledgments

I would like to thank my Ph.D. advisor, Matt Dwyer, for guiding me throughout my research career. Thanks also go to Sarfraz Khurshid and Willem Visser, with whom I started exploring symbolic execution techniques in the summer of 2002 and have continued a strong collaboration to this day. Antonio Filieri has been a great collaborator on pushing symbolic execution toward probabilistic reasoning, using model counting over symbolic constraints. Tevfik Butan and Pasquale Malacaria, and their students, have been wonderful collaborators on side channel analysis. My post doctoral researchers, Kasper Luckow, Rody Kersten, and Quoc-Sang Phan, have also made major contributions to the complexity analysis and side-channel analysis reported in this book. I would like to thank the many students and collaborators with whom I worked over the years on topics reported in this book. Finally, I would like to thank my family, Ada and Ionut, for their warm support.

Corina S. Păsăreanu
May 2020

CHAPTER 1

Introduction

In today's world, software systems have become more pervasive and complex leading to an increased need for techniques and tools that can ensure the safety and security of such systems. Testing is the most commonly used techniques for finding errors and vulnerabilities in software. However, it is typically a costly, manual process that accounts for a large fraction of software development and maintenance costs and often fails to uncover subtle, low-probability errors. Symbolic execution is a program analysis technique that can be used to automate software testing by systematically generating test inputs that can exercise (or "cover") large portions of the program and can thus help finding many subtle program errors.

The main idea behind symbolic execution [12, 26] is pretty straightforward. Rather than execute a program on specific, concrete inputs, as normally done when executing a program in a computer, symbolic execution attempts to execute a program on *symbolic inputs*, representing *multiple* concrete inputs all at once. When doing so, the values of program variables become mathematical expressions over the symbolic inputs. Because the symbolic inputs represent multiple concrete inputs, symbolic execution ends up following multiple program paths. For each path executed through the program, the analysis maintains a symbolic *path condition* which encodes the conditions on the inputs for the execution to follow that path. This path condition is built by accumulating the branch conditions encountered during the execution of the program. Whenever the path condition is updated it is checked for satisfiability using an off-the-shelf constraint solver. If the path condition is satisfiable it means that indeed there are concrete inputs that could make the program execute that path and the analysis continues; otherwise the analysis stops for that path, since no inputs can be generated to trigger that execution (in other words, the analysis backtracks). Test input generation is performed by solving the collected path conditions using the constraint solver. The obtained solutions represent the test inputs that follow different paths through the program. Symbolic execution can also be used directly for finding software bugs, where it checks for runtime errors or assertion violations during execution and it generates test inputs that trigger those errors.

The technique has been proposed in the 1970s but only more recently has symbolic execution become widely used, with many tools available, targeting different programming languages. This relatively recent resurgence is due to an increased availability of computation power, which makes a systematic analysis of program paths possible in practice, and to the rapid developments in the area of constraint solving [5, 14] leading to the rise of effective symbolic execution-based program analysis, testing and bug finding techniques.

Scalability is still the main limiting challenge for symbolic execution, due to the large (possibly infinite) number of paths to be explored and the complex constraints to be solved for most realistic software applications. Recent combinations of symbolic execution with random testing (or "fuzzing") have resulted in more scalable program analysis techniques which have achieved impressive results [8, 11, 21, 45]. For instance, in a recent DARPA Cyber Grand Challenge [13], a competition to create automatic defensive systems capable of reasoning about software flaws, formulating patches and deploying them on a network in real time, the 1st- [3] and 3rd-place [45] teams used techniques that combine symbolic execution and fuzzing in clever ways, for fully automated vulnerability detection and repair. While symbolic execution is routinely applied to test input generation and error detection, the technique has many other useful applications, such as regression analysis, load testing, robustness checking, or invariant generation [30, 36, 42, 48], to name a few.

In this book I give an overview of symbolic execution and some of its challenges (Chapter 2) and then focus on novel, less-studied, applications of symbolic execution to checking *non-functional* properties of software. The reviewed techniques investigate how symbolic execution can be used to check the *operation* of a system, rather than specific behavior. In particular, I first describe how symbolic execution can be used to finding *algorithmic complexity vulnerabilities* and other *performance* issues in software (Chapter 3). I further elaborate on a recent extension of symbolic execution with *probabilistic reasoning* (Chapter 4). This approach computes the probability of executing program paths by estimating the number of solutions for the collected path conditions and dividing that by the input domain. These path probabilities can be used to compute the *reliability* of a software component, which may be placed in different contexts, described by *probabilistic usage profiles*, and to compute the *information leakage* in side-channel analysis (Chapter 5). A chapter on directions for future research concludes this book.

CHAPTER 2

Symbolic Execution: The Basics

In this chapter, I give a gentle introduction to symbolic execution, using the simple example illustrated in Figure 2.1. On the left-hand side of the figure, there is some code that takes two integer inputs x and y and it checks x is greater than y, in which case some computation is performed to swap the two inputs. After this computation, x is checked again to see if it is greater than y, in which case an assert violation will happen. If the code swapping works correctly, this assert violation should not be possible. Testing this program for assert violations would typically involve assigning some concrete values to the inputs (say $x = 1$ and $y = 2$) and executing the code. As a result, only *one* program path will be executed; that does not lead to the error.

Instead of using concrete inputs, symbolic execution starts with symbolic values, which are denoted here as X and Y, and it initializes a *path condition*, PC, to true. Intuitively, the path condition for each path that is explored in the program maintains the conditions that the inputs need to satisfy in order for the execution to follow that path.

The result of symbolically executing the program is a *set of program paths* that can be conveniently organized into a *symbolic execution tree*, as illustrated in Figure 2.1 (right). The tree nodes represent program states and the tree edges represent program transitions between states. A symbolic state encodes the symbolic values and expressions for the program variables, the path condition, and the program counter. The tree represents, in a succinct form, *all* the possible executions through the code.

At each *if* statement, the path condition PC is updated with conditions on the input in order for the execution to choose between alternate paths. For example, at the execution of the first *if* statement, the condition could be either true or false; consequently, the PC is updated with the *if* condition or its negation, respectively, and symbolic execution proceeds with the systematic analysis

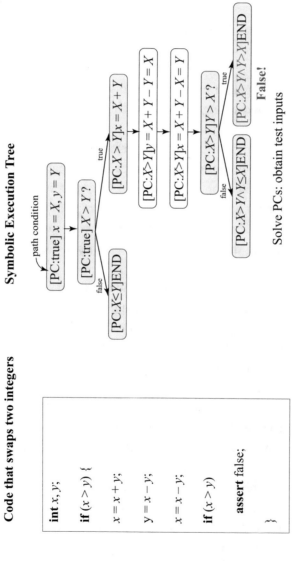

Figure 2.1: Example illustrating symbolic execution.

of both branches. This exploration can be performed in a depth-first, breadth-first, or some other heuristic search. If the path condition becomes unsatisfiable it means that the symbolic state is unreachable and symbolic execution stops for that path. Assignment statements assign symbolic expression to program variables, in an obvious manner.

For the example program, the analysis finds that the *assert* violation is unreachable; in general, this may not be the case because of overflow or underflow errors, but we chose to ignore such issues here, for simplicity.

Collection all the path conditions at the end of each symbolic path and solving them using a decision procedure, such as Z3 [14], results in solutions can be used as *test inputs* that are guaranteed to execute *all* the possible paths through the program.

2.1   HANDLING LOOPING PROGRAMS

Running symbolic execution on programs that contain loops or recursion results in a symbolic execution tree that is potentially *unbounded*. This problem is illustrated with the example in Figure 2.2. On the left-hand side of the figure, one can see the code of a simple method that takes as input integer n and uses it as the bound in a simple *while* loop. Note that the loop finishes for every concrete value of input n. Yet, the symbolic execution of this code produces an infinite symbolic execution tree, as illustrated in the figure, on the right-hand side. The reason is that the termination condition of the loop depends on the symbolic input (denoted S in this case), which is unconstrained. The repeated evaluation of the condition, corresponding to increasing the number of loop iterations, results in an unbounded number of new constraints, i.e., $S > 0$ and $S \leq 0$, $S > 1$ and $S \leq 1$, $S > 2$ and $S \leq 2$ and so on.

To deal with such situations, symbolic execution tools typically put a bound on the exploration depth, allowing them to terminate the analysis even in the case of programs with loops or recursion. In that case, the analysis will loose its *completeness* meaning that it will not be able to analyze all the possible paths through the program. However, it is still useful in practice as it may find errors that occur up to the specified bound. One can use loop invariants or abstraction technique to recover the completeness of symbolic execution; however these techniques are beyond this book.

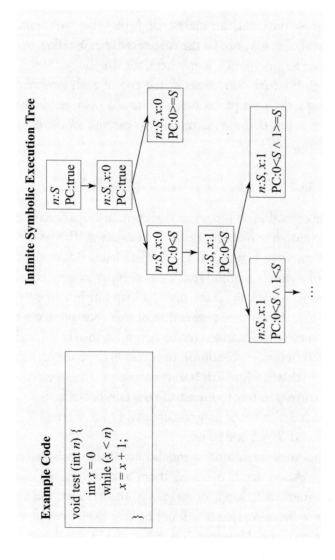

Figure 2.2: Example looping program.

2.2 DYNAMIC SYMBOLIC EXECUTION

In the previous section, I discussed some simple examples that illustrate "classic" symbolic execution which is essentially a static analysis technique, meaning that it analyzes a program without running it. Dynamic symbolic execution, on the other hand, collects symbolic constraints at *run time* during concrete executions. Examples include DART (Directed Automated Random Testing) [19] and Concolic (Concrete Symbolic) testing [43]. As dynamic execution is very simple to implement and very popular in practice, I am presenting it here in more detail.

A first version of the technique was actually proposed by Korel [28], and it consists of running the program starting with some random inputs, gathering the symbolic constraints on inputs at conditional statements, using a constraint solver to generate new test inputs and repeating the process until a specific program path or statement is reached. Both DART and concolic testing perform a similar dynamic test generation, where the process is repeated to attempt to cover *all* feasible program paths. During testing, these tools attempt to detect crashes, assert violations, and other runtime errors.

Specifically, dynamic symbolic execution maintains a concrete state and a symbolic state simultaneously and, for each collected path constraint, it systematically, or heuristically, negates one conjunct at a time. Off-the-shelf decision procedures or constraint solvers are invoked to *solve* the newly obtained constraints and the solutions are used as concrete test inputs, to steer the execution on alternative paths. This process is repeated until all feasible execution paths are explored or user-defined coverage criteria are met.

To illustrate dynamic symbolic execution, consider again the swap example that we presented before. Assume the analysis starts with $x = 0$, $y = 0$. At the same time the analysis creates two symbolic variables (denoted here also as x and y); the analysis collects the path constraints from the conditions encountered in the code during concrete execution. When executing the program on these inputs, the *else* branch of the first *if* condition is executed and the path constraint becomes $x \leq y$ (see Figure 2.3). This constraint is negated (it becomes $x > y$) and solved, obtaining $x = 1$ and $y = 0$. The program is run again on these new inputs. This time the *then* branch of the first *if* statement is executed. The value of x becomes 0 and the value of y becomes 1, i.e., the two values are swapped as intended. Correspondingly, the symbolic values become

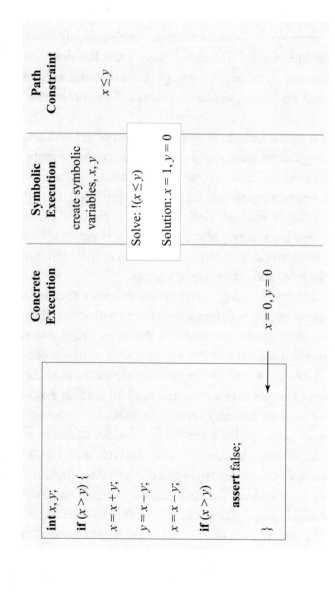

Figure 2.3: Dynamic symbolic execution for input $x = 0$, $y = 0$.

swapped as well; $x = y$ and $y = x$. The *else* branch of the second *if* statement is executed too and the path constraint becomes $x > y \land y \leq x$. Negating the last constraint results in a new path constraint $x > y \land y > x$ which is unsatisfiable; the analysis ends, with two paths explored; see Figure 2.4.

Thus, the dynamic approach explores the same number of paths as the static variant. The dynamic approach has several advantages over the static one, namely that it is easy to implement (e.g., via code instrumentation to collect the constraints) and that the concrete execution can be used to help the symbolic execution in certain situations, for instance when there are no available decision procedures or in the presence of native calls. The drawback with the approach is that it is based on re-execution, which can be expensive if the paths are very long, while the static approach can employ backtracking and can thus leverage incremental solving, which could be more efficient. Hybrid approaches, that combine static and dynamic analysis, are also possible. To illustrate the power of dynamic symbolic execution, consider the code below:

```
void test(int x, int y) {
    int z = x*x;
    if(y==z)
        assert false;
}
```

The code contains an operation on integer values that is nonlinear (multiplication) and reasoning about nonlinear integer constraints is in general undecidable. While it is possible to use constraint solvers to solve such constraints using bit-vector theory [14], assume one only has decision procedures that can reason about integer linear constraints. Thus, applying classical, static symbolic execution to this program is not possible, as the solver would not be able to reason about the path conditions generated from the execution of the *if* statement. On the other hand, dynamic symbolic execution can deal with this situation as follows. Suppose that the analysis starts with inputs $x = 1$ and $y = 2$. The concrete value of z is 1 and the symbolic value is $z = x * x$. Since z is different than y, the *else* branch of the *if* statement is executed, and the path condition becomes $y \neq x * x$, which is nonlinear and therefore the decision procedure cannot handle it. Instead of taking the symbolic value $z = x * x$ in the path condition, dynamic symbolic execution can take advantage of the fact

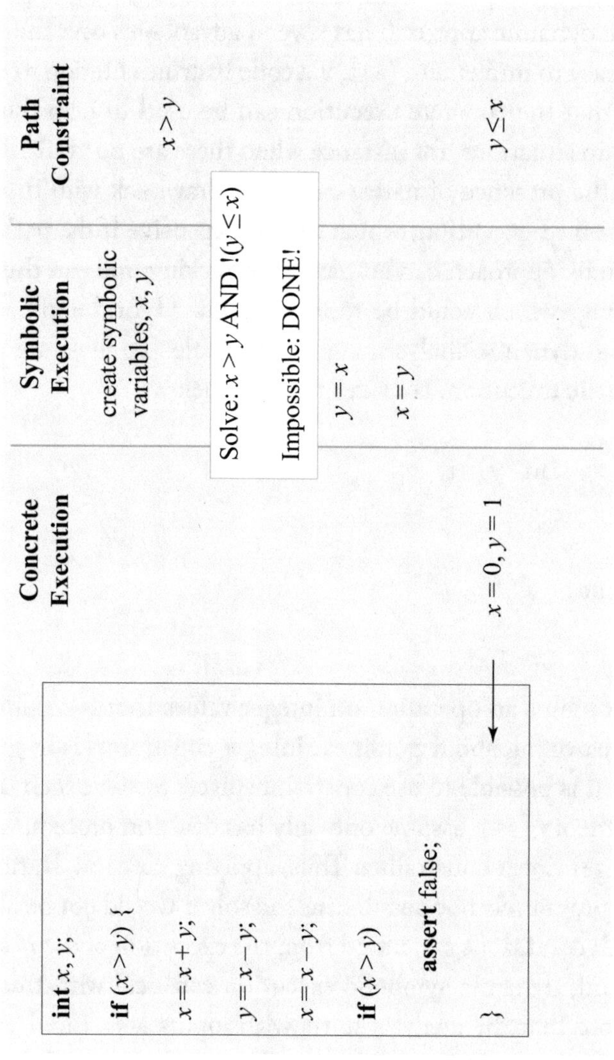

Concrete Execution	Symbolic Execution	Path Constraint
	create symbolic variables, x, y	
	Solve: $x > y$ AND $!(y \leq x)$ Impossible: DONE!	$x > y$
	$y = x$	
	$x = y$	$y \leq x$
$x = 0, y = 1$		

```
int x, y;

if (x > y) {

    x = x + y;

    y = x − y;

    x = x − y;

    if (x > y)

        assert false;

}
```

Figure 2.4: Dynamic symbolic execution for input $x = 1$, $y = 0$.

that the concrete values are available during execution, and use that to simplify constraints. For this simple example, one can use the concrete value $z = 1$ and simplify the path condition, which becomes $Y \neq 1$; the execution continues until the end of the procedure. To obtain inputs that guide the execution toward the *then* branch, one can then negate the constraint, obtain $y = 1$, and solve it, which can be done easily with the available decision procedure. The program is then re-executed with the new inputs: $x = 1$ and $y = 1$, the *then* branch is executed and the assert error is discovered. Thus, the dynamic symbolic execution tool is able to handle situations that a static tool can not.

Assume now that the second statement in the code is `int z = f(x);`, where `f` is some native library function that cannot be analyzed or instrumented directly. A static analysis tool would need to provide a model for such a function, while in a dynamic analysis tool, the same reasoning as above can be applied, therefore eliminating the need for explicit modeling of f. There may still be some situations, however, when such an approach would not work, e.g., due to side-effects of f. In that case, modeling f may be unavoidable.

Please note that dynamic symbolic execution faces the same challenges as its static counterpart in the presence of loops and recursive procedures, i.e., it may result in an unbounded number of paths. For this reason, a user-defined bound is still needed to ensure termination of the analysis.

2.3　HANDLING STRUCTURED INPUT

One of the challenges in symbolic execution, both static and dynamic variants, is the analysis of programs that take as inputs complex data structures, such as arrays, lists, or trees. The problem is difficult since the size and shape of the input is not known a-priori, and the method of collecting and solving mathematical constraints, as we have discussed in the previous sections, is no longer applicable. One way to address the problem is to use a technique that is called *lazy initialization*, first introduced in [25]. To perform symbolic execution over a function (or method) that processes complex inputs, the techniques starts with the concrete execution of the method under analysis on an input structure with *uninitialized* fields; these fields are then initialized with concrete information "lazily," when they are first accessed during the method's symbolic execution.

More precisely, consider a method m with one input object, the implicit input `this` (consider for a moment a Java-like semantics but the technique applies also to programs written in other languages). Methods with more than one parameter are treated similarly. To symbolically execute method m of class C, the technique first creates a new object, o, of class C and it sets all its fields to uninitialized values. The analysis then invokes o.m() and the execution proceeds following Java semantics for operations on reference fields and following traditional symbolic execution for operations on numeric fields, with the exception of the special treatment of (first) access to uninitialized fields, as illustrated in the following pseudo-code.

```
if (f is uninitialized) {
  if (f is reference field of class T) {
    nondeterministically initialize f to:
      - null
      - a new object of type T (with uninitialized field values)
      - a previously initialized object of class T
  }
  if (f is numeric (string) field)
    initialize f to a new symbolic value
}
```

In the procedure, whenever the symbolic execution needs to access a field f in the input data structure, it first checks if it is initialized or not. If it is not and f is a reference field, the procedure assigns it nondeterministically to either null, a new object of same type (with initialized fields) or a previously initialized object of the same class. Here, nondeterminism means systematic exploration of all the possibilities, a feature that is common in many program analysis tools. More operations need to be performed in the case of sub-typing, as lazy initialization should consider creating objects of all the valid sub-types, but the details are omitted here for simplicity. Thus, an uninitialized reference field is systematically initialized to different input values, according to different aliasing scenarios that may be possible in the input data structure. The underlying symbolic execution tool will systematically explore all the different scenarios, according to this procedure. If the field is of numeric or string type, the field gets assigned a fresh symbolic value, and the symbolic execution proceeds as for numeric inputs, described in the previous sections.

Consider now a concrete example, as illustrated in Figure 2.5 (left). The

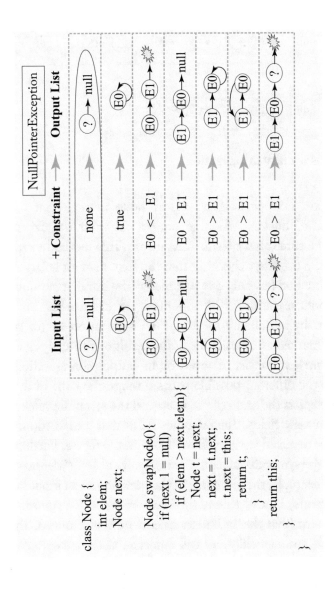

Figure 2.5: Illustration of lazy initialization on another swap example.

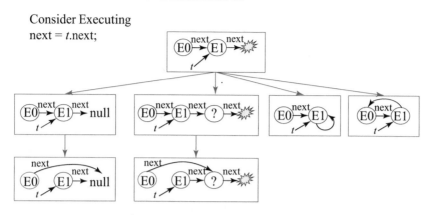

Figure 2.6: Details for lazy initialization.

figure shows a Java class Node {int elem; Node next;} that implements singly linked lists of integers. Assume one would want to analyze method swapNode, to make sure it has no run-time errors. This method swaps the first two elements in the list, after checking that the first element is larger than the second one. The method has only one parameter (the implicit parameter this). To perform symbolic execution on this method, one can use lazy initialization. The analysis starts by initializing the input list with an object (of type Node) whose nodes are marked as uninitialized. The result of the analysis is depicted in Figure 2.5 (right). It consist of seven input-output configurations of lists, corresponding to the different possible paths through the code of the method. A configuration depicts the *shape* of the heap and the symbolic values and constraints for the numeric fields. The reference fields that remained uninitialized after execution are depicted with a yellow "blob" in the figure. Furthermore, E_0 and E_1 represent the symbolic values of the integer fields; a "?" denotes an uninitialized field. For example, the last entry in the table shows an input list that has at least three elements, E_0 and E_1 and furthermore $E_0 > E_1$. Assume now that the first if statement (that checks if next in non-null) is removed. The analysis will check the code automatically and will report a NullPointerException for the first configuration in the table.

The exact details of lazy initialization are illustrated in Figure 2.6. Consider the execution of statement next = t.next in method swapNode. The configuration of the symbolic heap right before this statement is shown in the root of the tree. The symbolic heap resembles the concrete heap of the program except that some reference fields are uninitialized. Before executing the statement, the `next` field is uninitialized, meaning that it has not been accessed before by the symbolic execution along the path reaching this statement.

Since `t.next` is found to be uninitialized, the analysis first initializes it to: `null`, a new symbolic object (with uninitialized fields) or an object created during a previous initialization, resulting for this example in two circular lists. Intuitively, this means that the analysis makes four different assumptions about the shape of the input list according to different aliasing possibilities allowed by the declaration of class `Node`; the analysis explores all of them systematically. Once `t.next` has been initialized, the execution proceeds according to the standard Java semantics. Thus, nondeterminism is used to handle aliasing and the symbolic execution's search algorithm is used to explore all the different heap configurations explicitly. A similar lazy algorithm can be used for handling input arrays of unspecified size and content.

2.4 A PRACTICAL APPLICATION

I close this chapter with the presentation of a practical application from the aerospace domain. I will come back to this application later in this book, to illustrate reliability analysis based on symbolic execution (Chapter 4).

This application of symbolic execution was first described in [41], where a Java model of the prototype ascent abort handling software for NASA's Crew Exploration Vehicle (pictured in Figure 2.7) was analyzed using the Symbolic PathFinder tool. This piece of software is called the Onboard Abort Executive (OAE) and it was developed at NASA's Johnson Space Center (JSC). The OAE monitors the status of the vehicle during the ascent phase of flight and it decides when an abort is required, which abort mode is the safest for the astronauts, and when to automatically initiate an abort. The high-level structure of the code is shown in Figure 2.8. The OAE receives its inputs (e.g., current altitude, launch vehicle internal pressures, etc.) from sensors and other software components. The inputs are analyzed to determine if any of the ascent flight rules have been broken, and to evaluate which ascent abort modes are currently possible. If multiple abort modes are possible, the OAE chooses the abort mode that is safest for the flight crew. Once a flight rule is broken, and an abort mode is chosen, it is sent to the rest of the flight software for initiation. The analyzed code is approximately 600 lines of code, it has a large input space (approxi-

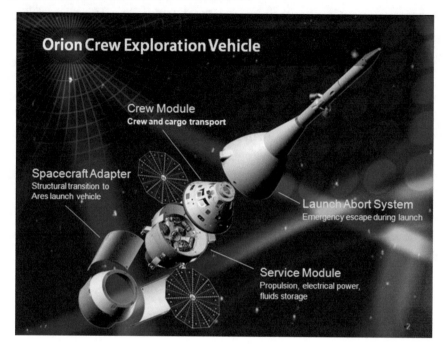

Figure 2.7: NASA's crew exploration vehicle.

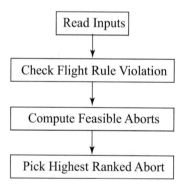

Figure 2.8: The onboard abort executive (OAE) structure.

mately 65 input variables) and fairly complicated logic. There are many ascent flight rules and abort modes, and encoding them correctly is difficult. Since the OAE is directly concerned with human safety, it is very important to have this code correctly implemented and carefully tested. At the time of this application, test generation was done by hand by the engineers at JSC, which was very time-consuming and provided little assurance that the OAE behaves as expected. The goal of the experiment in [41] was to determine symbolic execution can quickly and automatically generate all required test cases for the OAE. During test case generation, symbolic execution was also used to check for functional properties of OAE, which were derived from the documented requirements.

Example properties include: (1) if a flight rule is violated, then an abort mode must be chosen; (2) if no flight rules are violated, then an abort mode must not be chosen (OAE should output "no-abort"); and (3) if both performance and systems abort conditions exist, then the systems abort rules should apply. All these properties were encoded as assertions in the code.

The OAE required different kinds of code coverage for its test suite, as mandated by the developers at JSC. These include abort coverage, flight rule coverage, combinations of abort/flight rules coverage, and branch coverage. In a first phase of the experiment, symbolic execution was used to generate a test suite that included test cases violating multiple flight rules at the same time. However, since this was an early prototype, it was not designed to handle multiple flight rule violations happening at the same time. The test generation was therefore restricted to single flight rule violations (by instructing SPF to backtrack as soon as a second flight rule violation happened).

Symbolic execution generated approximately 200 test cases to cover all aborts and flight rules in less than a minute. During test case generation, an error was also discovered, highlighting a case where some flight rules were broken but no abort was chosen; this error was later corrected by the OAE developers. Each test case includes the values of the input variables and the expected output abort mode. Note that manual test case generation took more than 20 hours, did not cover all possible flight rule-abort combinations and missed the error. Furthermore, simple random testing managed to cover only a few flight rules and no aborts; this is not surprising since the OAE has a large input state space and complicated logic.

The engineers at JSC used the test cases that that were automatically generated both as part of their test suite for this early version of OAE and as regression tests for later versions. This proved to be important as the generated test cases identified a significant design error in a subsequent version of OAE and resulted in design changes that affected not just the flight rules and abort code, but also several other modules.

Note also that, in order to analyze the OAE, a driver was needed, that invokes the component on symbolic inputs (mimicking all the possible values coming from the sensors). This driver initially specified all the OAE inputs to be unconstrained, symbolic values. However, the OAE operates in a vehicle subject to real-world constraints, so some combinations of input parameters that might cause the code to detect flight rule violations are not actually possible in the real world. For example, the test suite generated with unconstrained inputs contained a case that set the inertial velocity of the vehicle to 24,000 ft/s, while the altitude of the vehicle was only 0 ft (a physically impossible combination, since the spacecraft can reach that velocity only at high altitude). Therefore, these constraints needed to be encoded explicitly to avoid generation of un-realistic test cases. The constraints were expressed as range restrictions on input variables or as relations over the variables. These were asserted as pre-conditions to the OAE code and were thereby automatically included in the path conditions. Some of the input constraints were directly provided by the domain experts, while others, such as range restrictions, were determined from simulation runs. Using these constraints, symbolic execution generated a test suite that did not contain physically impossible test cases, but still achieved the desired coverage.

Symbolic execution was applied to a component of a flight software system and it generated a test suite that obtained full testing coverage, for a special coverage required by the developer. The generation took a few seconds. In contrast, random testing obtained only partial coverage while manual test case generation took approximately 20 hours and was still not able to achieve adequate coverage. The analysis led to the discovery of critical errors that were later fixed by the developers.

CHAPTER 3

Symbolic Complexity Analysis

In this chapter I will talk about the application of symbolic execution to the analysis of non-functional properties of software systems with an emphasis on the analysis of worst-case algorithmic complexity in programs. The presentation follows [29]. The problem is to identify the program inputs that lead to the *worst-case* execution time or memory consumed by a program. Understanding such worst-case executions is important, as it can reveal performance bottlenecks in the program and it can also point out opportunities for program optimizations. Also of serious concern are *security vulnerabilities* that are due to algorithmic complexity, allowing a malicious user to easily build a "small" input that causes the system to consume an impractically large amount of resources (time or memory). By exploiting these vulnerabilities an adversary can mount Denial-of-Service (DoS) attacks in order to deny service to the system's benign users or to disable the system. Algorithmic complexity vulnerabilities are often the consequence of the algorithms used rather than of traditional "software bugs," and consequently traditional software bug hunting techniques are of little use to address them. Profilers can be used for finding performance bottlenecks in software, however they are inherently limited by the quality and number of test inputs used during profiling and by the platform on which the profiling was done.

Symbolic execution is a promising technique for finding algorithmic complexity vulnerabilities as it can systematically explore the program behaviors, and try to identify the particular inputs that lead to large resource (time or memory) consumption. Given a program component that takes inputs of a specified size, the technique from [29], dubbed SPF-WCA, computes the input constraints together with actual test input values that trigger the worst-case complexity of the program. In order to compute the (time/memory) consumption of each path, the analysis uses an abstract *cost model* which gives the cost (representing time or memory), for the execution of each instruction in the program.

Note that a naive symbolic execution of the program would need to explore every single path in the program and can thus not scale to realistic programs. To address this issue, SPF-WCA employs *path guidance policies* to efficiently search for worst-case behaviors. The policies are *learned* from the worst-case paths obtained with exhaustive symbolic execution of the program at small input sizes. The learned policies are then applied to guide the exploration of the program at larger input sizes. The intuition is that the worst-case paths obtained at small input sizes often follow the same decisions or sets of decisions when executing the conditional statements in the code. Further, the same decisions are "likely" followed by the worst-case paths at larger input sizes. Path policies encode succinctly in a data structure the decisions taken by symbolic execution along worst-case paths.

In [29] it is shown experimentally that the history-based policies are often precise enough to decide a unique choice at all decisions. Symbolic execution, guided by these policies, effectively reduces to exploring a single path regardless of input size and it scales far beyond the capabilities of non-guided, traditional symbolic execution for which the number of paths grows exponentially.

To get insight into the worst-case program behavior for *any* input size, SPF-WCA also uses techniques to *predict* the worst-case behavior based on the data obtained with the guided analysis. Specifically, an off-the-shelf library is used to *fit* a function to the data to obtain an estimate of the complexity as a function of the input size. The results of the function evaluation on increased input sizes are plotted to give insights to the developers into the worst-case complexity of the program. Comparing the estimates to the programmers' expectations or to theoretical asymptotic bounds can reveal vulnerabilities or confirm that a program's performance scales as expected.

A distinguishing feature of the work compared to previous similar techniques [10] is that the analysis employs path policies that take into account the *history* of choices made along the path to decide which branch to execute for the conditional statements in the program. For increased precision, the history computation is *context-preserving*, meaning that the decision for each conditional statement depends on the history computed with respect to the enclosing method.

3.1   APPROACH

The approach is illustrated in Figure 3.1. The analysis has two phases.

- *Policy Generation.* SPF-WC uses guidance policies for the *efficient* exploration of the symbolic execution tree of the analyzed program. This is in contrast to an exhaustive analysis, that, although guaranteed to find the worst-case path, would not scale due to the exponential number of paths that need to be explored and the cost of constraint solving. Intuitively, the guidance policies encode the decisions in the control flow graph that the symbolic execution needs to follow to find "likely" complexity vulnerabilities. A set of policies is first generated from an exhaustive symbolic execution for a range of small input sizes, where exhaustive exploration is still possible and the computation of the worst-case paths is precise. For each input size in the range, SPF-WCA obtains a policy that encodes the decisions taken along the respective worst-case path, as mapped on the CFG of the program.

- *Policy-Guided Exploration.* The obtained policy is used to *guide* the symbolic execution at larger input sizes. Thus, the approach is *guided* toward the paths that maximize the cost in terms of the cost model. All other choices not part of the policy are *pruned* from the search space, enabling a scalable exploration in many cases.

Finally, the worst-case costs collected for each input size are used for regression analysis to yield the characterization of the worst-case complexity of the program as a *function* of input size. In particular, SPF-WCA relies on multiple linear regression using the Ordinary Least Squares (OLS) method: it computes a function that minimizes the differences of the original data set and the predicted data points from the function.

SPF-WCA also provide a *theoretical guarantee* stating that when policies are obtained by taking the *union* of previously computed policies for increasing input sizes, the worst-case path policy is eventually found. More details can be found in [29].

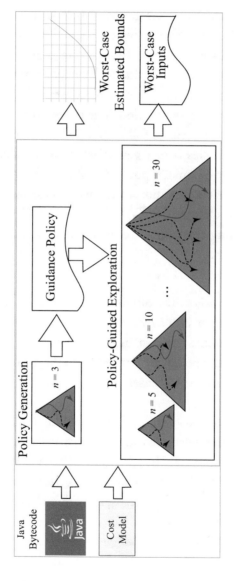

Figure 3.1: Policy generation and policy-guided exploration in SPF-WCA.

3.2   EXAMPLE

To illustrate the approach in more detail, consider the example below, which is taken from an interactive application that extracts commands from stdin and allows interaction with a backing store mapping commands to actions.

```
class Entry {
  String key; Action val; Entry next;
  public Entry(String key, Action val, Entry next) {
    this.key = key; this.val = val; this.next = next;
  }
}
Entry findEntry(String o, int n) {
  for(Entry e = table[n]; e != null; e = e.next) {
    if(e.key.equals(o)) {
      return e;
    }
  }
  return null;
}
class String {
  char[] val;
  // ...
  public boolean equals(Object oObj) {
    // ...
    String o = (String) oObj;
    if(val.length == o.val.length) {
      for(int i = 0; i < val.length; i++) {
        if(val[i] != o.val[i])
          return false;
      }
      return true;
    }
    return false;
  }
}
```

A malicious user could exploit the structure of the backing store by constructing inputs in such a way that all commands are stored in the same bucket, yielding hash collisions. Hash collisions are organized in a list, and by carefully crafting the inputs, the worst-case behavior can be exercised.

Note that to find an item in the linked list, the `findEntry` method iterates over it and compares the key strings to its o parameter. The worst-case behavior is not only determined by the structure of the hash table, which leads to worst-case linear look-up time instead of the usual constant time for hash tables, but also by the values of the keys that are stored in the table. Keys are compared with the `equals` method, which performs a character-wise comparison. Consequently, the worst-case behavior of the application happens when a non-existent key is looked up *and* that, when comparing all the entries, the characters in the keys are all equal *except* for the last character, leading to a worst-case execution of the `equals` method. This example is typical of applications that use fixed-sized strings (or codes) for keys. For strings of length k, the worst-case execution time for `findEntry` is $n \times k$, where n is the size of the input list.

Symbolic execution applied to `findEntry` for a small input list (where all the keys are symbolic) can already reveal this worst-case behavior. The worst-case path follows the edges marked in bold on the control flow graph (CFG) of method `equals`, as illustrated in Figure 3.2 (for simplicity only the CFG of `equals` is shown). Note that for condition $c : val[i] \neq o.val[i]$ (corresponding to the second `if` statement in method `equals`) in the graph (marked with grey) both *true* and *false* branches are taken along the worst-case path. This is because, as already discussed, that condition must be true to stay as long as possible in the loop, but it should be false when comparing the last character in the input string, so that method `equals` returns false, thus leading to the worst-case behavior in method `findEntry`. Also note that, for input size n and string length k, exhaustive analysis needs to explore k^n paths, which is infeasible for large input sizes n.

To obtain the worst-case behavior for large input sizes SPF-WCA attempts to *learn* a guidance policy to be used during the search for such paths. Specifically, the worst-case paths discovered with exhaustive exploration at small input sizes are used to derive a guidance policy that dictates which branch are *likely* to be taken during symbolic execution to exercise the worst-case path *for arbitrary input sizes*. However, a branch policy that simply dictates which

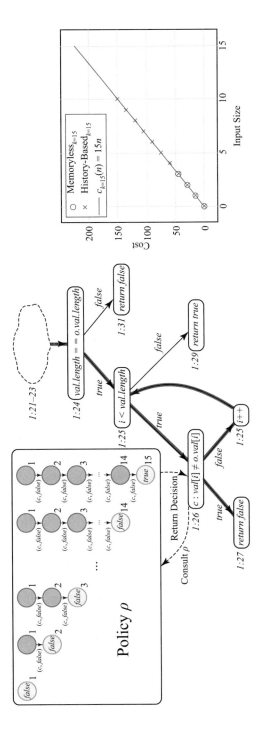

Figure 3.2: Policy generation for the Example.

choices to *always* make (as in [10]) when executing condition c would be insufficient for this example: the policy would prescribe *both* branches so no savings would be achieved. Thus, in this case, using a simple memoryless policy would be equivalent to exhaustive analysis and the guidance policy would not be useful.

To address this problem, SPF-WCA employs a more sophisticated guidance policy that takes into account the *history* of decisions taken along the worst-case path. For key size $k = 15$, the *false* branch must be taken at the decision on line 23 when comparing the characters for index $0, 1, \ldots, 13$ of the string; the *true* branch must be selected for index 14. Note that this history records the choices along a worst-case path during only one call of method `equals`, but the same pattern is followed for each invocation of method `equals` coming from `findEntry`. SPF-WCA therefore restricts the computation and storage of histories only with respect to a *calling context*. This has two advantages: first, it increases the precision of the analysis, since it can compute different histories for the same condition under different contexts; second, the histories are small since they are local to a particular context.

SPF-WCA computes automatically the *history-based* policy based on an exhaustive exploration at small input sizes ($n = 2$ in this case). The histories used to define the policy are encoded efficiently in a trie data structure that offers good compression: in this example, all 15 policies can be stored as a single path in a trie, because they are all prefixes of the longest decision history.

The computed policy is then used to guide the exploration at larger input sizes according to the pattern yielding the worst-case behavior. For the example, when condition c is evaluated during the guided symbolic execution, the symbolic path leading to it is examined to see if the previous up to 13 choices on c (the only symbolic condition in the current context) were all *false*, in which case the policy dictates that the *false* branch should be taken for the current condition (i.e., the *else* branch in the code). If, on the other hand, all the previous 14 choices were *false*, the policy dictates to take the *true* branch instead (i.e., the *then* branch in the code). This ensures that the worst case is exhibited for the `equals` and `findEntry` methods.

The guided search reduces to exploring a single path for an arbitrary number of elements added to the linked list data structure. SPF-WCA can thus expose the worst-case behavior and the concrete inputs that will exercise it. Moreover, SPF-WCA attempts to compute a model (function), mapping input sizes to observed worst case, using regression analysis. This model can then be used to predict the worst-case behavior for large (arbitrary) input sizes; see Figure 3.2 (right). The cost model for this example was computing the number of decisions along an execution path. Note that the history-based policy enables analysis at much larger inputs compared to memoryless guidance (or exhaustive analysis). Further, it enables obtaining concrete inputs that trigger the worst-case behavior at large input sizes, and building a more precise prediction model due to the larger data set.

CHAPTER 4

Probabilistic Reasoning

Symbolic execution typically involves exploring program paths and checking *qualitative properties* along each program path, leading to increased code coverage and to the discovery of assert violations and other run-time errors that may be present in the analyzed program. Symbolic execution can also be used to find worst-case behavior as discussed in the previous chapter. Let us study techniques that aim to not only find such paths but to also quantify *how many* paths satisfy certain properties. More precisely, the goal of these techniques is to estimate *how likely* it is for a program to satisfy given properties.

In recent work, symbolic execution has been extended with probabilistic reasoning [6, 17, 18], allowing one to reason about *quantitative* properties for software systems. The main idea is to estimate the *number of solutions* for the constraints computed with symbolic execution, and to use that information to compute the *probability* of target events of interest, such as the occurrence of an assertion in the code, a particular output or some other observable for the program, under an *input distribution*. Input distributions allow data from real-world observations to be incorporated in the analysis of programs that interact with their environment, as well as to encode uncertainty in design assumptions about the usage profile of a program, including the interactions with third-party components and systems. Computing probabilistic properties of software is useful in many domains including debugging, cryptography, biology, and reliability and security analysis.

In its simplest form, the analysis works as follows. Suppose one wishes to compute the probability of a particular event e in the program. Assuming a uniform input distribution (i.e., all inputs are equally likely) one can apply symbolic execution, and collect all the path conditions for the paths that lead to e. Assume for simplicity that the program has a finite number of paths (this assumption will be relaxed later in this chapter).

One can then *estimate* (or *count*) the number of input values that satisfy the path conditions and divide that by the size of the input domain (denoted

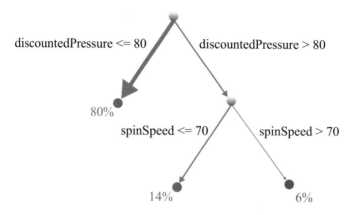

Figure 4.1: Simple example illustrating probabilistic reasoning.

$\sharp D$). Thus, the probability of event e is:

$$Pr(e) = \sum_i \frac{\sharp PC_i}{\sharp D}.$$

The counting can be done with off-the-shelf counting procedures, such as Latte [2] or Barvinok [1] for linear constraints.

4.1   EXAMPLE

As a simple example, consider the (fragment of) symbolic execution tree, taken from the OAE code, in Figure 4.1. Suppose the red state denotes some failure event. The path condition leading to it is *PC*: *spinSpeed* > 70 ∧ *discountedPressure* > 80. Assuming the size of the input domain is 100 × 100, one can compute the probability of *failure* as follows:

$$Pr(Fail) = \frac{\sharp PC}{\sharp D} = \frac{30 \cdot 20}{100 \cdot 100} = 6\%.$$

The figure also illustrates the probabilities computed for the other paths in the program, where the thickness of the drawn lines correlates with the probability value.

4.2   SOFTWARE RELIABILITY

One immediate application of the probabilistic reasoning defined above is software reliability analysis, which tackles the problem of predicting the failure probability of software, under prescribed usage scenarios. The term reliability is used here to refer to the probability of the software to successfully accomplish its assigned task when requested by the user.

Recall that a path condition is a set of constraints on the inputs that, if satisfied by the input values, will allow the execution to follow the specific path through the code. Assume that each (terminating) execution path is labeled with either *success* or *failure*. Since the set of path conditions produced by symbolic execution is a complete partition of the input domain, given a probability distribution on the inputs one can compute the reliability of the software as the probability of satisfying any of the *successful* path conditions. The probability distribution on the input formalizes the expected *usage profile* that accounts for all the external interactions of the software, both with the user and with external resources. However, remember that to account for non-termination in presence of loops, the analysis uses *bounded symbolic execution*. In this case, for the interrupted paths, it is not clear if the analysis will return success or failure; these paths are therefore treated specially, and labeled as *grey*. For an input satisfying a grey path condition, one cannot predict success nor failure. This *uncertainty* is used to define a confidence measure to assess the impact of the execution bounds on the reliability prediction.

4.3   APPROACH

In more detail, reliability analysis can be performed as follows. Suppose the program under analysis is symbolically executed using an off-the-shelf symbolic execution tool; the result is a set of symbolic paths, each with a path condition. Assume first that the symbolic execution of the program always terminates. The path conditions can be classified in two sets, according to the fact that they lead to success or failure, respectively:

$$\{PC_1^s, PC_2^s, \ldots, PC_m^s\}$$
$$\{PC_1^f, PC_2^f, \ldots, PC_p^f\}.$$

Note that all path conditions are disjoint by construction and cover the whole input domain.

Assume that all the input variables range over finite and countable domains, indicated generically as domain D. The *usage* of the program is described through a *probabilistic usage profile*, denoted UP. In its simplest form, UP can be defined as a set of pairs $\langle c_i, p_i \rangle$ where c_i is a *usage scenario*, defined as a (constraint representing a) subset of D and p_i ($p_i \geq 0$) is the probability that a user input belongs to c_i. Note that the $\{c_i\}$'s should define a complete partition of D, and thus $\sum_i p_i = 1$. Notice that c_i could contain even a single element of D, allowing for the finest grained specifications of UP.

One can then define the probability of successful termination for the program given the usage profile UP. This definition can be formalized as:

$$Pr^s(P) = \sum_i Pr\left(PC_i^s \mid UP\right).$$

Similarly, the failure probability $Pr^f(P)$ is defined as follows:

$$Pr^f(P) = \sum_i Pr\left(PC_i^f \mid UP\right).$$

Furthermore, $Pr^s(P) + Pr^f(P) = 1$ (since the path conditions partition the whole input domain of the program). The *reliability* of program P is then defined as the probability of success for program P:

$$Rel(P) = Pr^s(P).$$

These probabilities can be computed using model counting as follows. Since UP defines a partition of the input domain, it follows that:

$$Pr(PC \mid UP) = \sum_i Pr(PC \mid c_i) \cdot p_i.$$

Here PC is a path condition, c_i is one of the disjoint constraints in the usage profile, and p_i is the corresponding probability in the usage profile.

Furthermore, from the definition of conditional probability it follows that:

$$Pr(PC|c_i) = \frac{Pr(PC \wedge c_i)}{Pr(c_i)}.$$

In order to use model-counting techniques for the computation of the conditional probabilities, define for a constraint c the function $\sharp(c)$ that returns the number of elements of D satisfying c; it is always a finite, non-negative integer because D is finite and countable. It follows that $Pr(c)$ is $\sharp(c)/\sharp(D)$, where $\sharp(D)$ is the size of the domain that we implicitly assumed to be not null.

The probably of successful termination can be computed therefore as follows:

$$Pr^s(P) = \sum_i Pr\left(PC_i^s \mid UP\right)$$

$$= \sum_i \sum_j Pr\left(PC_i^s \mid c_j\right) \cdot p_j = \sum_i \sum_j \frac{\sharp\left(PC_i^s \wedge c_j\right)}{\sharp\left(c_j\right)} \cdot p_j.$$

The probability of failure is defined similarly.

4.4   GREY PATHS AND CONFIDENCE

As discussed, symbolic execution does not always terminate when the analyzed program has looping constructs and typically a bound is put on the exploration depth. In this setting the symbolic execution is no longer complete and, besides *success* and *failure* paths, a new set of paths is collected for the executions interrupted before reaching an error or completing the execution. These paths are labeled as *grey*, and have the corresponding path conditions:

$$\left\{PC_1^g, PC_2^g, \ldots, PC_q^g\right\}.$$

The probability of executing grey paths, $Pr^g(P)$, is computed similarly to the success or failure probabilities:

$$Pr^g(P) = \sum_i Pr\left(PC_i^g \mid UP\right).$$

The three sets $\{PC^s\}$, $\{PC^f\}$, and $\{PC^g\}$ are disjoint, and constitute a complete partition of the entire domain D. Therefore, the following holds:

$$Pr^s(P) + Pr^f(P) + Pr^g(P) = 1.$$

Note that $Pr^g(P)$ quantifies the inputs that lead to neither success nor failure at the current exploration depth. This information can be used to compute a measure of the confidence one can put on the reliability estimation, defined as follows:

$$Confidence = 1 - Pr^g(P).$$

A high confidence value indicates that the computed reliability is close to the actual one. *Confidence* = 1 means that the symbolic execution is complete.

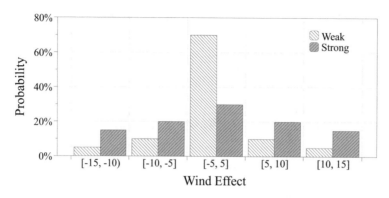

Figure 4.2: Example usage profiles.

However, a small value confidence value indicates that the exploration bound is too small and needs to be increased to get a better estimate of the actual reliability of the analyzed program.

4.5 USAGE PROFILES

The probabilistic analysis is defined with respect to a probabilistic usage profile that prescribes how the component should be used. The usage profile is meant to summarize succinctly hundreds of hours of operation or simulation of the actual environment of the component. Machine learning techniques can be used to build such usage profiles automatically. In this book, arbitrary usage profiles are handled through discretization. As an example, consider the usage profiles illustrated in Figure 4.2 capturing the effect of the wind, used for the reliability analysis for an aircraft controller in [17]. The effect of wind is uncertain and is profiled through a random variable resembling specific operation scenarios. The figure shows two profiles, corresponding to *weak* wind and *strong* wind, respectively. The wind effect ranges are reported on the x-axis. The vertical bars represent the probability of a value in the respective range to be used as input. Weak wind is more concentrated around zero, while strong is more likely to produce extreme values. Probabilistic analysis based on symbolic execution as described in this chapter can be used to estimate the mission success probability of the controller for the aircraft under the two usage profiles, for the two different wind conditions.

4.6   RELIABILITY ANALYSIS FOR THE ONBOARD ABORT EXECUTIVE (OAE)

This section describes the reliability analysis applied to the OAE component, which was first introduced in Chapter 2. Recall that the OAE monitors the status of the vehicle during the ascent phase of flight and it checks a set of *flight rules* that are meant to be invariant during the ascent. Whenever a flight rule is violated the OAE decides that an abort is required and it selects the abort mode which is safest for the astronauts. The OAE receives its inputs (e.g., current altitude, launch vehicle internal pressures, etc.) from sensors and other software components. The analyzed code is approximately 600 lines of code; it has a large input space (36 input variables) and complicated logic. The component itself did not have any errors, but an interesting question is to determine the probability of reaching an abort or conversely the probability of no abort, corresponding to mission success.

The OAE component was analyzed using different usage profiles for the OAE. One profile was a simple uniform distribution of values for each variable within its domain. Gaussian distributions for different input variables such as *thrust* and *tank_pressure* were also considered. The analysis revealed that the reliability of the component is very high (> 0.9999999) so the actual chance of a mission abort is extremely small; furthermore, the code seems to be very robust with regards to the different profiles.

4.7   MODEL COUNTING FOR DATA STRUCTURES

Model counting is of central importance in quantitative reasoning powered by symbolic execution. Model counting is the problem of computing the number of solutions (or models) that satisfy a set of constraints.

Most work on probabilistic symbolic execution has focused primarily on integer values and used off-the-shelf model counting tools for computing the number of integer values within the volume of a convex polytope, e.g., Latte [2] or Barvinok [1], to compute probabilities. More recently, techniques have been proposed for solution space quantification of floating-point [6, 7] and string [4] constraints.

However these techniques cannot be directly applied to complex data representations, such as lists and trees. I describe here a model counting procedure for a combination of heap and numeric constraints obtained with a symbolic execution of a program written in a modern programming language such as Java. A simple approach is to enumerate all the possible data structures up to a given size and then to check their validity against the given constraints. However, this becomes quickly intractable for large solution sets. One can use instead the *lazy initialization* technique which was described earlier in this book (see Chapter 2). The approach was first described in [16] aims to generate and thus count data structures that satisfy a mixture of heap and numeric constraints. Lazy initialization is used the lazily enumerate the heap constraints while off-the-shelf model counters are used for the numeric constraints.

Recall that lazy initialization extends symbolic execution with the ability of handling input data structures: it constructs the heap as the program paths are explored, and defers concretization of symbolic heap objects as much as possible. Lazy initialization produces symbolic heaps that are pairwise non-isomorphic while guaranteeing that no relevant states are missed. It can thus be used as a powerful procedure for generating and *counting* all the structures (up to a given bound) that satisfy a set of constraints.

The heap constraints can have the following forms.

- $ref = null$. Reference ref points to *null*.

- $ref \neq null$. Reference ref is non *null*.

- $ref_1 = ref_2$. Reference ref_1 points to the same object in the heap as reference ref_2, i.e., ref_1 and ref_2 are aliased.

- $ref_1 \neq ref_2$. References ref_1 and ref_2 are not aliased.

4.8   EXAMPLE

The approach is illustrated on the following code; this is the same `swapExample` we discussed earlier in the book (see Chapter 2).

```
class Node {
    int elem;
    Node next;

    Node swapNode() {
```

```
    if(elem > next.elem) {
      Node t = next;
      next = t.next;
      t.next = this;
      return t;
    }
    return this;
  }
}
```

Symbolic execution with lazy initialization results in seven symbolic paths, due to the if condition and the different aliasing possibilities in the input. The resulting heap constraints and numerical constraints can be encoded as logical formulas as follows.

PC_1: *in.next* $=$ *null* $\land$ *in* $\neq$ *null*

PC_2: *in.next* $=$ *in* $\land$ *in* $\neq$ *null*

PC_3: *in.next* $\neq$ *in* $\land$ *in.next* $\neq$ *null* $\land$ *in* $\neq$ *null* $\land$ *in.elem* $\leq$ *in.next.elem*

PC_4: *in.next.next* $=$ *null* $\land$ *in.next* $\neq$ *in* $\land$ *in.next* $\neq$ *null* $\land$ *in* $\neq$ *null* $\land$ *in.elem* $>$ *in.next.elem*

PC_5: *in.next.next* $=$ *in* $\land$ *in.next* $\neq$ *in* $\land$ *in.next* $\neq$ *null* $\land$ *in* $\neq$ *null* $\land$ *in.elem* $>$ *in.next.elem*

PC_6: *in.next.next* $=$ *in.next* $\land$ *in.next* $\neq$ *in* $\land$ *in.next* $\neq$ *null* $\land$ *in* $\neq$ *null* $\land$ *in.elem* $>$ *in.next.elem*

PC_7: *in.next.next* $\neq$ *in* $\land$ *in.next.next* $\neq$ *in.next* $\land$ *in.next.next* $\neq$ *null* $\land$ *in.next* $\neq$ *in* $\land$ *in.next* $\neq$ *null* $\land$ *in* $\neq$ *null* $\land$ *in.elem* $>$ *in.next.elem*

For example, PC_3 encodes an input list (denoted by *in*) whose first two elements are not aliased (*in.next* $\neq$ *in*), they are not *null* (*in.next* $\neq$ *null* $\land$ *in* $\neq$ *null*), and furthermore, the first element is smaller than or equal to the second element in the list (*in.elem* $\leq$ *in.next.elem*).

The symbolic paths represent all the possible actual executions of swapNode and the PCs represent an isomorphism partition of the input space. There is one *failure* path, for PC_1—the method raises an un-handled `NullPointerException`. There are no grey paths (since there are no loops). Thus, one can encode the constraints provided by symbolic execution together with the constraints from the usage profile as a special Java method returning a boolean value: true if the input data structure satisfies the constraints, false if it doesn't.

Applying symbolic execution with lazy initialization over this special method will then enumerate all the valid data structures (up to a pre-specified bound). The valid assignments for the symbolic numeric fields can then be counted with a suitable model counting procedure, such as Latte.

The model counting procedure thus requires two inputs.

Special boolean method: A boolean method encoding the constraints; returns true if the structure satisfies the constraints (e.g., the list is acyclic).

Finitization: Domain bounds for both reference and numeric data (e.g., a list may have up to 5 nodes, whose elements are between 1 and 10).

For example, the predicate needed to count all the acyclic lists having at most 6 nodes, whose elements are between 1 and 10, is as follows.

```
class List{
        Node head;
        boolean PredAcyclic(){
                Set<Node> nodes = new HashSet<Node>();
                Node iterator = head;
                while(iterator!=null){
                        // check acyclic
                        if(!nodes.add(iterator))
                                return false;
                        //check bounds
                        if(iterator.elem<1||iterator.elem>10)
                                return false;
                        if(nodes.size>6)
                                return false;
                        iterator=iterator.next;
                }
```

```
                    return true;

           }

}
```

The symbolic execution of the method PredAcyclic collects as successful path conditions (i.e., not leading to false) all the symbolic structures representing an acyclic list with at most 6 nodes, whose elements are integers between 1 and 10. The total number of acyclic lists can thus be obtained applying established model counting solutions on the success path conditions, which now predicate only on the numeric fields. The result in this case would be 6,543,210 acyclic lists out of 7,654,321 lists with up to 6 nodes (and elements between 1 and 10).

To finish the original swapNode example, consider computing the probability of failure (in this case, throwing a NullPointerException). Assume a usage profile that specifies that the input list is acyclic with probability 0.9 and it is cyclic with remaining probability 0.1. There is only one failure symbolic path (revealed by a null pointer exception in the evaluation of the if condition). The path condition for the failure path is as follows:

$$input \neq null \wedge input.next = null.$$

Since this path condition is only satisfiable for acyclic lists, the probability of failure can be computed as follows:

$$Pr^f(P) = 0.9 \cdot \frac{\sharp(input \neq null \wedge input.next = null \wedge acyclic(input))}{\sharp(acyclic(input))}.$$

Model counting gives the following results: $\sharp(input \neq null \wedge input.next = null \wedge acyclic(input)) = 10$ and $\sharp(acyclic(input)) = 1,111,111$, for lists with up to 6 nodes and elem ranging over 1 ...10, giving probability of failure $8.1 \cdot 10^{-6}$. One can argue that one should simply correct the error in method swapNode (by adding a null check or a method precondition stating that the input list should not be null). However, a quantitative analysis like the one presented here, can be used to rank the discovered errors in a large code base, helping developers to focus on fixing the high-probability ones first.

CHAPTER 5

Side-Channel Analysis

In this chapter I touch upon the application of symbolic execution and probabilistic reasoning in the security domain, namely for the analysis of side channels in programs. Side channels enable a malicious user to recover secret program data from non-functional characteristics of computations, such as time or power consumption, number of memory accesses, or size of output files. Side-channel attacks have been shown to pose serious threats, e.g., by recovering cryptographic keys from the RSA encryption/decryption algorithm [9], and private information about users, as with commonly used algorithms for data compression [24]. There is thus an increased need for practical tools that can detect and prevent side-channel vulnerabilities.

Symbolic execution can be used for the automatic analysis of side-channels as documented in [32, 35, 37]. The analysis can compute *quantitative* bounds on the amount of information (i.e., number of bits in the secret) that can be leaked via side-channels. The idea is that the amount of leaked information can be correlated to the number of different possible side-channel observations. These observations can be quantified using symbolic execution and model counting. Similar to the complexity analysis described in Chapter 3, the side-channel analysis is parametrized by a *cost model* which gives the side-channel measurements (in terms of consumed time and memory, bytes written to a file, and so on) observed from the execution of program instructions. The side-channel "observations" are the values for the cost computed for each path in the analyzed program. The *number* of observations is estimated from the number of symbolic paths (and the model count for the constraints corresponding to each path).

In the following, I describe in more detail the side channel analysis for estimating the leakage. I further discuss how the extension of symbolic execution with *probabilistic reasoning* is used to compute information theoretic metrics for side-channel leakage, such as Shannon or Smith's min entropy [44].

Finally, I describe a method for automatic attack synthesis. This method derives the public user input that results into *maximum leakage*, mimicking the worst-case attacker. This public input is computed by using weighted MaxSMT solving [33], which is a generalization of SMT solving to optimization problems and it is used here to obtain the maximal assignment over the set of clauses obtained with symbolic execution.

The obtained solution gives a public input that leads to the largest number of possible observations, hence giving the maximal leakage: any other choice of public input would result in fewer observables. This public input is interesting since it gives an estimate of the worst-case "attack step" that a malicious user can perform in order to guess secret information in one step.

The approach generalizes naturally over multiple-step side-channel attacks where symbolic execution can be used to quantify the information revealed to an attacker after multiple channel measurements made. This corresponds to an attacker who makes multiple guesses by invoking and measuring the execution of the program multiple times on different public inputs to gradually uncover a secret that is constant across program runs (such as the secret key used in the RSA encryption/decryption algorithm). In this case, MaxSMT solving is used to compute a *sequence* of public inputs that lead to maximum leakage, exposing the vulnerability of the program to multi-run attacks.

5.1   EXAMPLE

To illustrate side-channels and the problems they may pose, consider the following simple password checking example.

```
boolean verifyPassword(byte [] input, byte [] password) {
  for ( int i = 0; i < SIZE; i++) {
   if (password[i] != input[i])
     return false ;
   Thread.sleep(25L);
  }
  return true;
}
```

As it is the convention in the security literature, assume that the input data is partitioned in public vs. private data. The public, non-sensitive input can be directly accessed and manipulated by an attacker, while the private, sensitive input can not.

Method `verifyPassword` takes a *public* value (`input`) and a `private` value (*password*) and compares them element by element; the method returns false as soon as a mismatching value is found. The method is insecure to an attacker who can measure the time taken by the method to return true or false. This is because the method leaks some information about the password when the inputs do not match and the method will have a different execution time based on the length of the common prefix of the secret and the public information.

If the attacker were only allowed to observe the result (true or false) returned by the password checking method, the attacker would need a number of tries that is exponential in the length of the password to guess the full password. However, if the attacker can measure the execution time for the method, she would need only a linear number of tries to guess the password. The attacker will first try to guess systematically the first element in the password; she will know when the guess is correct when the timing will change. She will then move to guessing the next element and so on. Thus, the attacker needs $O(n \times m)$ tries where n is the length of the password and m is the size of the element.

Note that, for simplicity, the assumption here was that the attacker is able to make *perfect* timing observations. In practice, multiple measurements per observation are necessary to account for noise in the execution environment.

5.2 QUANTITATIVE INFORMATION FLOW ANALYSIS

Quantitative Information Flow (QIF) is a powerful approach to *"measure"* leaks of confidential information in a software system. Typically simple, qualitative, information flow analysis accepts programs as secure if confidential data cannot be inferred by an attacker through their observations of the system—this intuitive property is called *non-interference*. Although satisfying non-interference is a sound guarantee for a system to be secure, this requirement is too strict for most realistic programs as to leak some information is almost unavoidable. By quantifying leakage QIF addresses this limitation: QIF accepts as secure programs with "small" interference, not just with "zero interference" (non-interference).

A fundamental QIF result (the channel capacity theorem [31, 44]) states that leakage for a program is always less than or equal to the log of the number of possible distinct observations that an attacker can make. By noting these observables with (*NObs*) and channel capacity with *CC*, the channel capacity theorem states the following:

$$\text{Information leaked} \leq \log_2(NObs) = CC(P).$$

This result means that, in essence, QIF reduces to *counting* the number of different observable outputs for the program. The result holds for different notions of leakage based on the probability of guessing the secret or the notion of leakage based on Shannon's information theory measuring the number of bits leaked. For these reasons counting the number of observables is the basis of state-of-the-art QIF analysis [15, 23, 27, 38–40].

The Channel Capacity can be computed using symbolic execution in a straightforward way. One can run symbolic execution to collect all symbolic paths of the program, where both public and private inputs are symbolic. Assuming that the analyzed program is terminating and deterministic, and that the input domain is finite, it follows that the program has only a finite number of symbolic paths. For simplicity, assume that all the paths terminate within the prescribed bound. If this is not the case, one can use the notion of *confidence* described in Chapter 4.

Each symbolic path will have a cost, according to a cost model. The number of observations can be computed by *counting* the number of paths that have different cost. Let $\mathcal{O} = \{o_1, o_2, ...\}$, be the set of possible costs. Note that different symbolic paths may have the same cost but a path cannot have more than one cost. The Channel Capacity, i.e., the maximal possible leakage is then computed as follows:

$$CC(P) = \log_2(|\mathcal{O}|).$$

5.3   SHANNON ENTROPY

There are other information theory measures, such as Shannon entropy $\mathcal{H}(P)$, that can be used to compute bounds for the leakage, when the secret distribution is known.

The Shannon entropy is defined as follows, where $p(o_i)$ is the probability for each observation o_i:

$$\mathcal{H}(P) = - \sum_{i=1,m} p(o_i) \log_2 (p(o_i)).$$

For deterministic systems, the Shannon entropy can also be used to give a measure of the leakage of the side-channel, corresponding also to the observation gain (on the secret) after one round of observation.

The entropy $\mathcal{H}(P)$ can be computed using symbolic execution and its probabilistic extension, which was already discussed in Chapter 4.

Let $\mathcal{O} = \{o_1, o_2, ...\}$ be the set of possible observations as defined before. Let π_j denote a symbolic path, and PC_j the corresponding path condition. Let us assume for simplicity a uniform distribution over the secret. The probability of observing o_i is then as follows:

$$p(o_i) = \frac{\sum\limits_{cost(\pi_j)=o_i} \sharp(PC_j(h,l))}{\sharp D}.$$

Here $\sharp(PC_j(h,l))$ denotes the number of solutions (i.e., secret and public concrete values) of constraint $PC_j(h,l)$ and $\#D$ denotes the size of the input domain D assumed to be (possible very large but) finite. The number of solutions can be computed with an off-the-shelf procedure such as Latte or Barvinok as before.

5.4   EXAMPLE

Going back to the password checking example and assuming a 4-bit input password, $D = 256$; symbolic execution of method `verifyPassword` results in five paths:

- $h[0] \neq l[0]$ returns false: 128 values

- $h[0] = l[0] \wedge h[1] \neq l[1]$ returns false: 64 values

- $h[0] = l[0] \wedge h[1] = l[1] \wedge h[2] \neq l[2]$ returns false: 32 values

- $h[0] = l[0] \wedge h[1] = l[1] \wedge h[2] = l[2] \wedge h[3] \neq l[3]$ returns false: 16 values

- $h[0] = l[0] \wedge h[1] = l[1] \wedge h[2] = l[2] \wedge h[3] = l[3]$ returns true: 16 values

When the observable is execution time, $H = 1.875$; when the observable is the returned output, $H = 0.33729$, which is much smaller, indicating a much smaller leakage.

The code for the corrected password checking is provided below:

```
boolean verifyPassword(byte [] input, byte [] password){
  boolean matched=true;
  for(int i = 0; i < SIZE; i++) {
   if (password[i]!=input[i])
     matched=false ;
   else
     matched=matched;
   Thread.sleep(25L);
  }
  return matched;
}
```

This example no longer has the side-channel vulnerability, as all the paths have (approximately) the same execution time.

5.5   ATTACK SYNTHESIS

An interesting problem to study is to determine the public inputs that trigger the maximum leakage of secret values in a program. Solving this problem allows one to obtain a precise quantification of the leakage and also provides a demonstration to the developers of an intuitive *attack* that can recover secret information, convincing them of the validity of the analysis.

To tackle the problem, first remark that symbolic execution of a program where both public and secret inputs are symbolic may result in an over-approximation in the computed bounds. The problem is illustrated by the following toy example:

```
void example(int lo, int hi) {
 if(lo<0){
  if(hi<0) Thread.sleep(10L);
  else if(hi<5) Thread.sleep(20L);
  else Thread.sleep(30L);
 }
 else {
  if(hi>1) Thread.sleep(40L);
  else Thread.sleep(50L);
 }
}
```

Here lo (low) denotes a public input while hi (high) denotes the secret. There are five possible observables, corresponding to five different timings, for this program, yielding a leakage of $\log_2(5)$ according to the channel capacity theorem. However note that there is actually no public input that can lead to this large leakage in a single program run. Indeed, if lo has a negative value, there are only three observations possible (i.e., different costs), while if lo has a positive value, only two possible observations can be made. Thus, the maximum number of possible observations after one round of observations made by an attacker can only be three, giving maximum leakage of only $\log_2(3)$.

The *goal* of the attack synthesis is to find the public input that leads to a maximum number of observables, as this value will also maximize the leakage (using the channel capacity formulation). Intuitively, this public value would show the most vulnerable behavior of the program and would characterize the most powerful *attack* in one step. The analysis can be further extended to compute the input that maximizes the Shannon entropy (see [37]).

This public value can be computed using MaxSMT solving as follows. Recall that MaxSMT is a generalization of SMT to optimization. Given a set of weighted clauses, MaxSMT finds the solution that maximizes the *sum* of the weights of the satisfied clauses. The MaxSMT problem is formulated by building a clause for each cost, where the clause is the disjunction of the path conditions that lead to the same cost, and the weight is 1. Intuitively, each clause defines all the secrets that lead to the same cost. The MaxSMT solver is then used to pick the public value that satisfies the most clauses, meaning that it allows the attacker to gather the maximum information about the secret, after one round of observations. By construction, any other public value will lead to fewer observations, allowing the attacker to infer fewer information about the secret.

To illustrate, for the example above, the following clauses are built, each with weight 1:

$C_1 :: (l < 0 \wedge h_1 < 0)$
$C_2 :: (l < 0 \wedge h_2 \geq 0 \wedge h_2 < 5)$
$C_3 :: (l < 0 \wedge h_3 \geq 5)$
$C_4 :: (l \geq 0 \wedge h_4 > 1)$
$C_5 :: (l \geq 0 \wedge h_5 \leq 1)$.

Here l and h denote the symbolic values of the lo and hi inputs, respectively. Note that each path condition was renamed such that all the public symbolic values have symbolic fresh values h_i. Intuitively, the renamed path conditions define constraints on public values (while the private values are left unconstrained) and the goal is to find the public input value that leads to the maximum number of observations for any value of the secret. For the example, MaxSMT reports that C_1, C_2, C_3 are together satisfiable with $l = -1$ as a solution resulting in maximum weight 3. Thus, the public input that achieves maximum leakage can be automatically computed; it is $l = -1$ and the leakage is $\log_2(3) = 1.58$ bits.

5.6   MULTI-RUN SIDE-CHANNEL ANALYSIS

The side-channel analysis described in the previous section generalizes naturally to multiple program runs, corresponding to the case where the attacker learns the secret by observing multiple program executions. This can be achieved by analyzing the *composition* of the program:

$$P(h, l_1); P(h, l_2); \ldots P(h, l_k).$$

In other words, for a k-step attack we consider running the same program k times, with different symbolic inputs: $l_1, l_2, \ldots, l_k$. Note that h remains the same across the k runs (i.e., we assume the secret is the same across the runs). In this case an *observable* is a *sequence* of costs, one cost for each program run. MaxSMT can then be used to derive the *sequence* of low values that lead to the maximum number of observation sequences after k steps.

To illustrate the multi-run side channel anlysis, consider again the example above. In two runs, MaxSMT returns 4 satisfiable clauses out of 13 distinct clauses. Hence, there are 4 observables and the maximum leakage is 2 bits. The low inputs are $l_1 = 0$, for the first attack step, and $l_2 = -1$, for the second attack step.

Note that in practice, one does not need to apply symbolic execution over the composition of k copies of the program, which could be very expensive. It is sufficient to apply symbolic execution only once, collect all the path conditions and then assemble them in k combinations to correspond to the path conditions for the composed programs. Also note that in practice, a *greedy* approach for computing a multiple step attack can be used, corresponding to an attacker who always pick the low input that maximizes the leakage each step. This approach would not necessarily return the sequence with the maximal possible leakage at each step, but could scale well in practice. Other possible attackers can be modeled in a similar fashion, e.g., an adaptive attacker that chooses the next low input following the observables from the previous rounds. In this case the attacker can be modeled as a function from the sequence of observables (the history) to low inputs (the next input). This approach has been explored in [32] to compute adaptive attacks that maximize Shannon leakage, using parameterized model counting and numerical optimizations.

Going back to the password example, Figure 5.1 shows the results of computing the attack for a password with 4 elements with 4 values each (8 bits of information).

5.7   THE EFFECT OF MULTI-THREADING

So far, in this book, the analyzed programs have been assumed to be deterministic. In general, this may not be the case, as programs can contain multi-threading and calls to random functions, making the analysis more involved. I discuss here the effect of multi-threading (or more generally of non-determinism) on a side-channel analysis. One way to extend the analysis to multi-threaded programs is to use *maximal* linear schedules (i.e., thread interleavings) similar to [17]. An in-depth description of side-channel analysis for noisy programs (where the noise can come from randomness in the program and imprecise timing measurements) is given in [32]. Suppose there is no knowledge about the next-choice distribution or the specific scheduling policy for the thread scheduler. The goal

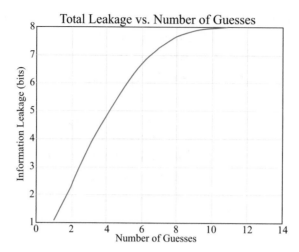

Figure 5.1: Computing a multi-run attack for password check.

is to identify the public inputs *and* the sequence of thread choices that is leading to the worst leakeage.

First note that each thread scheduling induces a sequential program on which one can apply the leakage analysis as described above. As there is a finite number of schedules for the current exploration depths, one can therefore compute the low input and maximum leakage for each one of them and report the worst case to the user.

The schedules produced by symbolic execution can be organized into a prefix tree. A schedule is defined to be *maximal* if it is not a prefix of any other schedule in the paths reported by a (bounded) symbolic execution.

For a thread schedule that is maximal, define a set of path conditions

$$\Pi_S = \{PC_i | S_i \in \textit{prefix} (S)\},$$

where *prefix(S)* is the set of all the prefixes of S including S. Intuitively, Π_S contains all the path conditions for the paths that "follow" the same schedule S, accounting for possible early termination.

For a maximal schedule the path conditions corresponding to Π_S cover the entire domain input. Thus, Π_S is the result of symbolically executing program $P(h, l)$ restricted to schedule S which can be seen as executing an equivalent deterministic program $P_S(h, l)$. One can therefore compute the value of the public (low) input that maximizes leakage by applying the methods described in the previous section (essentially the path conditions used for computing the leakage will be the path conditions corresponding to each path in Π_S). Maximal schedules can then be ordered according to their worst leakage and the worst result is reported to the user.

As an example, consider two threads:

T_1 :: example(lo,hi);
T_2 :: lo=-lo;

As before, hi and lo are private and public inputs, respectively. Method example is described in the code above. Assume for simplicity that thread 1 invokes method example atomically (e.g., in a synchronized block). Then lo=-1 gives 3 observables for thread schedule "T_1; T_2" but only 2 observables for "T_2; T_1." Thus, there may be more than one program execution associated with an input, each with different observations, and only some executions give the worst leakage.

Note that the number of possible thread interleavings may be very large and enumerating all of them may be infeasible. The problem can be addressed by partial order reduction (POR), supported by standard model checkers. POR exploits the commutativity of concurrently executed instructions, which result in the same state when executed in different orders.

Note also that the non-determinism introduced by multi-threading may produce some "noise" that makes the leakage smaller and this is not captured by the analysis presented here. However, if the attacker is allowed to observe the scheduling sequence, the leakage increases as in the example. The analysis produces a thread schedule which can be analyzed by developers for debugging and can thus be quite useful in practice.

CHAPTER 6

Conclusion and Directions for the Future

In this book I described symbolic execution and highlighted some of its applications to the analysis of programs, with the goal of ensuring their safety and security. These applications include software bug and vulnerability detection, complexity analysis, reliability analysis, and side-channel quantification and attack synthesis.

Scalability remains the main challenge for symbolic execution. Hybrid analysis techniques, that combine fuzzing with symbolic execution, are particularly promising in addressing this challenge. Fuzzing has emerged as one of the most promising techniques for finding bugs and security vulnerabilities in software, and it has been used routinely in industry, at large companies such as Microsoft [20] and Google [47]. For instance, the AFL fuzzing tool [47] was instrumental in finding Stagefright vulnerabilities in Android, Shellshock vulnerabilities in BIND, and numerous vulnerabilities in popular applications, such as OpenSSL, OpenSSH, Apache, or PhP.

Fuzzing performs essentially random testing with some guidance resulting into a cheap technique, that scales very well but may not be good at finding deep paths that depend on complicated constraints. Symbolic execution, on the other hand, is expensive but it is particularly good at finding deep paths. Thus, several researchers have observed that it is better to use these techniques together, leading to powerful tools for software analysis [3, 45].

To illustrate the power of such hybrid analysis techniques, consider Figure 6.1. It highlights some results from the application of a hybrid approach for worst-case analysis of a smart contract for crypto currency, taken from [34]. This is a self-executing contract with the terms of the agreement between buyer and seller being directly written into lines of code. Such smart contracts are typically metered, using Ethereum gas as internal pricing for running transactions. Exceeding allocated budget could result in loss of crypto currency. Cost

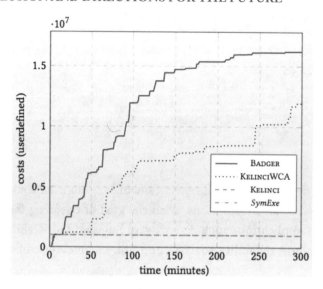

Figure 6.1: Analysis of a smart contract.

depends on concrete value in the input and can thus be very large (even for a short path). Analysis of worst-case gas consumption is illustrated in the figure. Badger is the name of the tool presented in [34]; it combines symbolic execution with fuzzing for worst-case analysis. Kelinci is a tool that performs fuzzing while KelinciWCA incorporates special heuristics for worst-case analysis. Note that the hybrid analysis performs much better than symbolic execution or fuzzing taken separately, indicating the promise of such hybrid techniques. Current hybrid techniques cannot perform well when the inputs to the application are highly structured or when the application contains randomness and/or concurrency and more research is needed to address these limitations.

Also worth mentioning is the increased interest in the exploration of symbolic execution, and in general of program analysis techniques, to the analysis of machine-learning methods, such as deep neural networks. These are computing systems, organized in multiple *layers* of *neurons*, that progressively learn to perform tasks by considering labeled examples. Symbolic execution can be applied to such systems in a straightforward way, by viewing the networks as imperative programs that apply repeated transformations to their inputs [22, 46]. However, scalability challenges are particularly acute for such networks, since they have a large number of paths and huge input spaces. Furthermore, neural networks are highly interconnected structures, making it difficult to develop exploration heuristics that are effective. New compositional and aggressive parallel techniques are needed to make symbolic execution of such systems practical.

Bibliography

[1] Barvinok library. http://garage.kotnet.org/skimo/barvinok/ 32, 37

[2] LattE. http://www.math.ucdavis.edu/latte/ 32, 37

[3] T. Avgerinos, D. Brumley, J. Davis, R. Goulden, T. Nighswander, A. Rebert, and N. Williamson. The mayhem cyber reasoning system. *IEEE Secur. Priv.*, 16(2):52–60, 2018. DOI: 10.1109/msp.2018.1870873 2, 55

[4] A. Aydin, W. Eiers, L. Bang, T. Brennan, M. Gavrilov, T. Bultan, and F. Yu. Parameterized model counting for string and numeric constraints. In G. T. Leavens, A. Garcia, and C. S. Păsăreanu, Eds., *Proc. of the ACM Joint Meeting on European Software Engineering Conference and Symposium on the Foundations of Software Engineering, ESEC/SIGSOFT FSE*, pp. 400–410, Lake Buena Vista, FL, November 4–9, 2018. DOI: 10.1145/3236024.3236064 37

[5] C. Barrett and C. Tinelli. CVC3. In *Proc. of the 19th International Conference on Computer Aided Verification, CAV'07*, pp. 298–302, Springer-Verlag, Berlin, Heidelberg, 2007. DOI: 10.1007/978-3-540-73368-3_34 2

[6] M. Borges, A. Filieri, M. d'Amorim, C. S. Păsăreanu, and W. Visser. Compositional solution space quantification for probabilistic software analysis. In M. F. P. O'Boyle and K. Pingali, Eds., *ACM SIGPLAN Conference on Programming Language Design and Implementation, PLDI'14*, pp.123–132, Edinburgh, UK, June 9–11, 2014. DOI: 10.1145/2594291.2594329 31, 37

[7] M. Borges, Q. Phan, A. Filieri, and C. S. Păsăreanu. Model-counting approaches for nonlinear numerical constraints. In C. W. Barrett, M. Davies, and T. Kahsai, Eds., *NASA Formal Methods—9th International Symposium, NFM Proceedings*, Moffett Field, CA, May 16–18, 2017, vol. 10227 of *Lecture Notes in Computer Science*, pp. 131–138, 2017. DOI: 10.1007/978-3-319-57288-8_9 37

[8] E. Bounimova, P. Godefroid, and D. A. Molnar. Billions and billions of constraints: Whitebox fuzz testing in production. In *35th International Conference on Software Engineering, ICSE'13*, pp. 122–131, San Francisco, CA, May 18–26, 2013. DOI: 10.1109/icse.2013.6606558 2

[9] D. Brumley and D. Boneh. Remote Timing Attacks are Practical. In *Proc. of the 12th Conference on USENIX Security Symposium—Volume 12, SSYM'03*, vol. 1, USENIX Association, Berkeley, CA, 2003. DOI: 10.1016/j.comnet.2005.01.010 43

[10] J. Burnim, S. Juvekar, and K. Sen. WISE: Automated test generation for worst-case complexity. In *IEEE 31st International Conference on Software Engineering*, pp. 463–473, May 2009. DOI: 10.1109/icse.2009.5070545 22, 28

[11] S. K. Cha, T. Avgerinos, A. Rebert, and D. Brumley. Unleashing mayhem on binary code. In *IEEE Symposium on Security and Privacy, SP*, pp. 380–394, San Francisco, CA, May 21–23, 2012. DOI: 10.1109/sp.2012.31 2

[12] L. A. Clarke. A program testing system. In *Proc. of the Annual Conference*, pp. 488–491, 1976. DOI: 10.1145/800191.805647 1

[13] Darpa cyber grand challenge (CGC). https://www.darpa.mil/program/cyber-grand-challenge 2

[14] L. De Moura and N. Bjørner. Zvol.3: An efficient SMT solver. In *Proc. of the 14th International Conference on Tools and Algorithms for the Construction and Analysis of Systems, TACAS'08*, pp. 337–340, Springer-Verlag, Berlin, Heidelberg, 2008. DOI: 10.1007/978-3-540-78800-3_24 2, 7, 11

[15] G. Doychev, D. Feld, B. Köpf, L. Mauborgne, and J. Reineke. CacheAudit: A tool for the static analysis of cache side channels. In *Proc. of the 22nd USENIX Conference on Security, SEC'13*, pp. 431–446, USENIX Association, Berkeley, CA, 2013. DOI: 10.1145/2756550 46

[16] A. Filieri, M. F. Frias, C. S. Păsăreanu, and W. Visser. Model counting for complex data structures. In *Model Checking Software—22nd International Symposium, SPIN Proceedings*, pp. 222–241, Stellenbosch, South Africa, August 24–26, 2015. DOI: 10.1007/978-3-319-23404-5_15 38

[17] A. Filieri, C. S. Păsăreanu, and W. Visser. Reliability analysis in symbolic pathfinder. In D. Notkin, B. H. C. Cheng, and K. Pohl, Eds., *35th International Conference on Software Engineering, ICSE'13*, pp. 622–631, IEEE Computer Society, San Francisco, CA, May 18–26, 2013. DOI: 10.1109/icse.2013.6606608 31, 36, 51

[18] J. Geldenhuys, M. B. Dwyer, and W. Visser. Probabilistic symbolic execution. In M. P. E. Heimdahl and Z. Su, Eds., *International Symposium on Software Testing and Analysis, ISSTA*, pp. 166–176, ACM, Minneapolis, MN, July 15–20, 2012. DOI: 10.1145/2338965.2336773 31

[19] P. Godefroid, N. Klarlund, and K. Sen. DART: Directed automated random testing. In *PLDI'05*, June 2005. DOI: 10.1145/1064978.1065036 9

[20] P. Godefroid, M. Levin, and D. Molnar. Automated whitebox fuzz testing. In *NDSS'08*, February 2008. 55

[21] P. Godefroid, M. Y. Levin, and D. A. Molnar. SAGE: Whitebox fuzzing for security testing. *Commun. ACM*, 55(3):40–44, 2012. DOI: 10.1145/2090147.2094081 2

[22] D. Gopinath, M. Zhang, K. Wang, I. B. Kadron, C. S. Păsăreanu, and S. Khurshid. Symbolic execution for importance analysis and adversarial generation in neural networks. In *30th IEEE International Symposium on Software Reliability Engineering, ISSRE*, pp. 313–322, Berlin, Germany, October 28–31, 2019. DOI: 10.1109/issre.2019.00039 56

[23] J. Heusser and P. Malacaria. Quantifying information leaks in software. In *Proc. of the 26th Annual Computer Security Applications Conference, ACSAC'10*, pp. 261–269, ACM, New York, 2010. DOI: 10.1145/1920261.1920300 46

[24] J. Kelsey. Compression and information leakage of plaintext. In *Revised Papers from the 9th International Workshop on Fast Software Encryption, FSE'02*, pp. 263–276, Springer-Verlag, London, UK, 2002. DOI: 10.1007/3-540-45661-9_21 43

[25] S. Khurshid, C. S. Păsăreanu, and W. Visser. Generalized symbolic execution for model checking and testing. In *Tools and Algorithms for the*

Construction and Analysis of Systems 9th International Conference, TACAS Proceedings, pp. 553–568, Warsaw, Poland, April 7–11, 2003. DOI: 10.1007/3-540-36577-x_40 13

[26] J. C. King. Symbolic execution and program testing. *Commun. ACM*, 19(7):385–394, July 1976. DOI: 10.1145/360248.360252 1

[27] B. Köpf, L. Mauborgne, and M. Ochoa. Automatic quantification of cache side-channels. In *Proc. of the 24th International Conference on Computer Aided Verification, CAV'12*, pp. 564–580, Springer-Verlag, Berlin, Heidelberg, 2012. DOI: 10.1007/978-3-642-31424-7_40 46

[28] B. Korel.   A dynamic approach of test data generation.   In *IEEE Conference on Software Maintenance*, November 1990. DOI: 10.1109/icsm.1990.131379 9

[29] K. Luckow, R. Kersten, and C. Păsăreanu. Symbolic complexity analysis using context-preserving histories. In *Proc. of the 10th IEEE International Conference on Software Testing, Verification and Validation (ICST)*, pp. 58–68, 2017. DOI: 10.1109/icst.2017.13 21, 22, 23

[30] R. Majumdar and I. Saha. Symbolic robustness analysis. In *IEEE Real-Time Systems Symposium*, pp. 355–363, 2009. DOI: 10.1109/rtss.2009.17 2

[31] P. Malacaria and H. Chen.   Lagrange multipliers and maximum information leakage in different observational models.   In *Proc. of the 3rd ACM SIGPLAN Workshop on Programming Languages and Analysis for Security, PLAS'08*, pp. 135–146, New York, 2008. DOI: 10.1145/1375696.1375713 46

[32] P. Malacaria, M. H. R. Khouzani, C. S. Păsăreanu, Q. Phan, and K. S. Luckow. Symbolic side-channel analysis for probabilistic programs. In *31st IEEE Computer Security Foundations Symposium, CSF*, pp. 313–327, IEEE Computer Society, Oxford, UK, July 9–12, 2018. DOI: 10.1109/csf.2018.00030 43, 51

[33] R. Nieuwenhuis and A. Oliveras. On SAT modulo theories and optimization problems. In *Proc. of the 9th International Conference on Theory*

and Applications of Satisfiability Testing, SAT'06, pp. 156–169, Springer-Verlag, Berlin, Heidelberg, 2006. DOI: 10.1007/11814948_18 44

[34] Y. Noller, R. Kersten, and C. S. Păsăreanu. Badger: Complexity analysis with fuzzing and symbolic execution. In *Proc. of the 27th ACM SIGSOFT International Symposium on Software Testing and Analysis, ISSTA*, pp. 322–332, Amsterdam, The Netherlands, July 16–21, 2018. DOI: 10.1145/3213846.3213868 55, 56

[35] C. S. Păsăreanu, Q. Phan, and P. Malacaria. Multi-run side-channel analysis using symbolic execution and MAX-SMT. In *IEEE 29th Computer Security Foundations Symposium, CSF*, pp. 387–400, IEEE Computer Society, Lisbon, Portugal, June 27–July 1, 2016. DOI: 10.1109/csf.2016.34 43

[36] S. Person, G. Yang, N. Rungta, and S. Khurshid. Directed incremental symbolic execution. In *PLDI'11*, (to appear), June 2011. DOI: 10.1145/2345156.1993558 2

[37] Q. Phan, L. Bang, C. S. Păsăreanu, P. Malacaria, and T. Bultan. Synthesis of adaptive side-channel attacks. In *30th IEEE Computer Security Foundations Symposium, CSF*, pp. 328–342, IEEE Computer Society, Santa Barbara, CA, August 21–25, 2017. DOI: 10.1109/csf.2017.8 43, 49

[38] Q.-S. Phan and P. Malacaria. Abstract model counting: A novel approach for quantification of information leaks. In *Proc. of the 9th ACM Symposium on Information, Computer and Communications Security, ASIA CCS'14*, pp. 283–292, New York, 2014. DOI: 10.1145/2590296.2590328 46

[39] Q.-S. Phan, P. Malacaria, C. S. Păsăreanu, and M. d'Amorim. Quantifying information leaks using reliability analysis. In *Proc. of the International SPIN Symposium on Model Checking of Software, SPIN*, pp. 105–108, ACM, New York, 2014. DOI: 10.1145/2632362.2632367

[40] Q.-S. Phan, P. Malacaria, O. Tkachuk, and C. S. Păsăreanu. Symbolic quantitative information flow. *SIGSOFT Softw. Eng. Notes*, 37(6):1–5, November 2012. DOI: 10.1145/2382756.2382791 46

[41] C. Păsăreanu, P. Mehlitz, D. Bushnell, K. Gundy-Burlet, M. Lowry, S. Person, and M. Pape. Combining unit-level symbolic execution and

system-level concrete execution for testing NASA software. In *ISSTA'08*, July 2008. DOI: 10.1145/1390630.1390635 17, 19

[42] C. S. Păsăreanu and W. Visser. Verification of java programs using symbolic execution and invariant generation. In *SPIN*, pp. 164–181, 2004. DOI: 10.1007/978-3-540-24732-6_13 2

[43] K. Sen. Concolic testing. In *ASE'07*, ACM, New York, 2007. DOI: 10.1145/1321631.1321746 9

[44] G. Smith. On the foundations of quantitative information flow. In *Proc. of the 12th International Conference on Foundations of Software Science and Computational Structures, FOSSACS'09*, pp. 288–302, Springer-Verlag, Berlin, Heidelberg, 2009. DOI: 10.1007/978-3-642-00596-1_21 43, 46

[45] N. Stephens, J. Grosen, C. Salls, A. Dutcher, R. Wang, J. Corbetta, Y. Shoshitaishvili, C. Kruegel, and G. Vigna. Driller: Augmenting fuzzing through selective symbolic execution. In *23rd Annual Network and Distributed System Security Symposium, NDSS*, San Diego, CA, February 21–24, 2016. DOI: 10.14722/ndss.2016.23368 2, 55

[46] Y. Sun, M. Wu, W. Ruan, X. Huang, M. Kwiatkowska, and D. Kroening. Concolic testing for deep neural networks. In *Proc. of the 33rd ACM/IEEE International Conference on Automated Software Engineering, ASE*, pp. 109–119, Montpellier, France, September 3–7, 2018. DOI: 10.1145/3238147.3238172 56

[47] M. Zalewski. American fuzzy lop (AFL). http://lcamtuf.coredump.cx/afl/, 2017. 55

[48] P. Zhang, S. G. Elbaum, and M. B. Dwyer. Automatic generation of load tests. In *ASE*, pp. 43–52, 2011. DOI: 10.1109/ase.2011.6100093 2

Author's Biography

CORINA S. PĂSĂREANU

Corina S. Păsăreanu is an ACM distinguished scientist, working at NASA Ames Research Center. She is affiliated with Carnegie Mellon University's CyLab and holds a courtesy appointment in Electrical and Computer Engineering. At Ames, she is developing and extending Symbolic PathFinder, a symbolic execution tool for Java bytecode. Her research interests include model checking and automated testing, compositional verification, model-based development, probabilistic software analysis, software engineering for machine learning, autonomy, and security. She is the recipient of several awards, including ASE Most Influential Paper Award (2018), ESEC/FSE Test of Time Award (2018), ISSTA Retrospective Impact Paper Award (2018), ACM Impact Paper Award (2010), and ICSE 2010 Most Influential Paper Award (2010). She has been serving as Program or General Chair for several conferences including: FM 2021, ICST 2020, ISSTA 2020, ESEC/FSE 2018, CAV 2015, ISSTA 2014, ASE 2011, and NFM 2009. She is currently an associate editor for the *IEEE TSE Journal*.

Printed in the United States
by Baker & Taylor Publisher Services